# Summary and Analysis of

# THE SIGNAL AND THE NOISE

Why So Many Predictions Fail—
but Some Don't

Based on the Book by Nate Silver

**WORTH BOOKS**
SMART SUMMARIES

All rights reserved, including without limitation the right to reproduce this book or any portion thereof in any form or by any means, whether electronic or mechanical, now known or hereinafter invented, without the express written permission of the publisher.

This Worth Books book is based on the 2015 paperback edition of *The Signal and the Noise* by Nate Silver, published by Penguin Books.

Summary and analysis copyright © 2017 by Open Road Integrated Media, Inc.

ISBN: 978-1-5040-4674-9

Worth Books
180 Maiden Lane
Suite 8A
New York, NY 10038
www.worthbooks.com

Worth Books is a division of Open Road Integrated Media, Inc.

The summary and analysis in this book are meant to complement your reading experience and bring you closer to a great work of nonfiction. This book is not intended as a substitute for the work that it summarizes and analyzes, and it is not authorized, approved, licensed, or endorsed by the work's author or publisher. Worth Books makes no representations or warranties with respect to the accuracy or completeness of the contents of this book.

# Contents

| | |
|---|---|
| Context | 1 |
| Overview | 3 |
| Summary | 5 |
| Direct Quotes and Analysis | 39 |
| Trivia | 43 |
| What's That Word? | 47 |
| Critical Response | 53 |
| About Nate Silver | 55 |
| For Your Information | 57 |
| Bibliography | 59 |

# Context

In 2008, statistician and baseball analyst Nate Silver became a national figure when he correctly predicted forty-nine out of fifty states in the US presidential election on his website, FiveThirtyEight. By the time his sought-after first book, *The Signal and the Noise*, was published in September 2012, Silver was writing for the *New York Times* and had become the go-to guru of political forecasting in the United States. He has since relaunched FiveThirtyEight as a stand-alone website, owned by ESPN, that covers politics, sports, culture, and more.

As many became interested in the possibilities of big data and books that discussed the application of data to business, such as *Moneyball: The Art of Win-

**SUMMARY AND ANALYSIS**

*ning an Unfair Game* by Michael Lewis, Silver established himself as one of the foremost experts behind official numbers and polling predictions.

One of the most consistent themes of Silver's approach to forecasting has been a move away from the idea of punditry getting it "right" or "wrong" in predicting elections or other events. In this probability-based approach to forecasting, he runs thousands of possible scenarios through a model that determines how often a result is likely to happen.

In the final predictions for the 2016 presidential general election, Silver's model showed roughly a 70% likelihood of Hillary Clinton winning the election with Donald Trump winning the other 30%. So while the chances were higher that Clinton would win, it was not statistically impossible—nor even very unlikely—that Trump would succeed.

# Overview

*The Signal and the Noise* is Nate Silver's take on how data has been used in a variety of fields and how people in the era of big data should approach forecasting. In each chapter, Silver selects a different field—from politics to earthquakes to poker—and analyzes the strengths and weaknesses of its commonly used methods of prediction.

The book's title refers to the "signal," or a set of data points that indicates that something is going to happen, and the corresponding "noise," or other pieces of information that are often included that can make the signal hard to read. An easily understood example can be found in economic forecasting: There are almost always indicators of an impending finan-

## SUMMARY AND ANALYSIS

cial collapse, but they are often lost amid other distracting data points.

*The Signal and the Noise* examines the biases that are brought into many predictions, particularly in areas like economics and terrorism, and offers advice on how forecasters can deal with the problem of too little information. For instance, while weather forecasters have become increasingly skilled at predicting temperatures and precipitation given an abundance of information, predicting the spread of infectious diseases is complicated by poor information and lack of consistency in data.

Moving into methodology, Silver focuses on the Bayesian theory, a forecasting technique that calculates the probability of an event by assessing various existing factors and constantly adding new information in hopes that predictions will continuously improve. He returns again and again to this theory to explain why, for instance, the stock market is not infallible, as it is subject to the biases and imperfections of the humans who influence it.

Silver's broad thesis is that, using computer modeling and Bayesian theory, forecasters in a variety of fields can continue to improve their predictions. The hardest areas are ones that involve an unpredictable human element, such as terrorism, but with diligence and an awareness of personal bias, Silver maintains that we can develop progressively more accurate predictions.

# Summary

## Introduction

The invention of the printing press in 1440 led to the spread of information around Europe. Seventy-seven years later, the publication of Martin Luther's *Ninety-five Theses* plunged the continent into centuries of religious war. But this dissemination of ideas also led to the Industrial Revolution, which brought about extraordinary economic growth.

In the 2000s, big data has made huge progress in areas from baseball to betting, but still, entire sectors, from financial catastrophes to natural disasters, are not being predicted accurately. The amount of data available makes it almost impossible for us to sort through it all, but if we begin to understand our natural biases, we can use the data to help us make better predictions.

**Need to Know:** Just as the printing press brought a wave of information that destabilized Europe but ultimately led to the Industrial Revolution, the rise of big data will bring a lot of bad information—noise—which we must learn to sift for the truth—the signal.

## 1. A Catastrophic Failure of Prediction

The financial crisis of 2008 was a result of a failure of prediction by the agencies who gave excellent ratings to securities known as collateralized debt obligations (CDOs), which included bad mortgages and were incredibly unsafe. The rating agencies profited from the abundance of CDOs, so they incentivized banks to continue producing more.

The unprecedented American housing bubble in the 2000s was result of people being encouraged to buy or flip homes even if they couldn't afford to do so. This was supported by a financial market—lenders, brokers, and ratings agencies—which benefited from every sale—and from the incorrect belief that homeownership was always a profitable investment.

For each dollar that was being spent in housing

sales, there were almost fifty dollars worth of trades in mortgage-backed securities. Institutions like Lehman Brothers were highly leveraged, and they were betting with money that they didn't actually have or that they had borrowed, which put them in precarious positions if the value of their portfolios declined even a small amount. This should have made investors reluctant to purchase assets, but the positive ratings from the credit agencies convinced unknowing buyers that these were solid purchases.

When Barack Obama took office in 2009, his stimulus package was meant to keep unemployment in check; in reality, the recession was worse than people knew at the time, causing the unemployment rate to go higher than his administration predicted it would.

The recession was caused by a series of poor predictions, each caused by predictors overlooking key pieces of information. In the lead-up to the housing crash, ratings agencies were creating models that didn't include all data relevant to the current housing situation, making them useless. An important lesson was learned the hard way. We should always take all data into account, even data that disrupts our models and calls our accuracy into question. A false sense of confidence in accuracy can lead to avoidable disaster.

**Need to Know:** The housing crisis and subsequent financial collapse and economic recession were caused

## SUMMARY AND ANALYSIS

by a string of prediction errors, each a result of overlooking key information.

## 2. Are You Smarter Than a Television Pundit?

In the run-up to the 2008 presidential election, many television pundits were unable to predict the obvious: Barack Obama had a solid lead and was going to win. With this knowledge, Silver goes back to evaluate the predictions made on the public affairs show *The McLaughlin Group* and determines that, overall, they only got about half of their forecasts right. A similar trend could be seen in the 1980s, when political experts failed to predict the collapse of the Soviet Union, despite Gorbachev's sincere efforts to reform the country and the dire economic straits within the USSR. Historically, forecasts from experts in a variety of subject areas were barely more accurate than random chance.

While studying the personalities of these experts, Philip Tetlock divided them into two categories—hedgehogs and foxes. Hedgehogs believe in big ideas and maintain that governing principles affect all behavior. Foxes believe in a plethora of small ideas and understand that there is nuance and complexity in the world.

Tetlock then found that the latter group is better at predicting. For instance, foxes would have been able to see that the USSR was an increasingly unstable country for many reasons, while hedgehogs saw only an "evil empire" or a socialist stronghold. But hedgehogs—with their bold, unwavering beliefs—make better TV guests. For hedgehogs, the more information they have, the more likely they are to twist the data to fit with their pre-held beliefs, missing or ignoring any information that would disrupt their forecasts.

FiveThirtyEight was founded by Silver's desire to approach the 2008 Democratic primary with qualitative analysis rather than cable news fluff. It was founded on three "fox-like" principles:

*Principle 1: Think Probabilistically*

The forecasts on FiveThirtyEight are probabilistic, meaning they cover a range of likely outcomes, accounting for real world uncertainty. What this means practically is that an event with a 90% chance of happening will still not happen 10% of the time. This doesn't mean the prediction was incorrect.

*Principle 2: Today's Forecast Is the First Forecast of the Rest of Your Life*

Probabilities are moving targets and will change as new data is considered each day. A key to

## SUMMARY AND ANALYSIS

FiveThirtyEight predictions is the willingness to change as information becomes available.

*Principle 3: Look for Consensus*

Hedgehogs want to single-handedly predict a major event and bask in the glory of their skills, but foxes realize that the best way to forecast is to aggregate many predictions and look for consensus in the data.

**Need to Know:** Confident political pundits (hedgehogs) are likely to predict inaccurately because they are blinded by their own biases, while data-driven forecasters (foxes) can combine many perspectives to see the truth more accurately and make better predictions.

## 3. All I Care About Is W's and L's

In the baseball prediction system Silver created for Baseball Prospectus, PECOTA (Player Empirical Comparison and Optimization Test Algorithm), Silver had determined that Red Sox player Dustin Pedroia would be a success, despite scout reports that dismissed him. Silver's predictions proved to be true, and when he sought an interview with Pedroia about these numbers, he realized that the key to the athlete's success was his above-it-all attitude. Pedroia hadn't listened to the scouting reports, which could have brought him down.

Baseball projection systems must account for the context of the available statistic, decipher skill from luck, and understand the "aging curve," or how a player's performance changes as he ages. The first task is easy enough, but the second and third are much harder. With PECOTA, Silver used similarity scores to compare current players to similar past players as a way to predict future success.

Ever since the publication of *Moneyball* in 2003, the divide between stat-focused teams and scouting-focused teams has shrunk, as both have appreciated the strengths of the other. Silver pitched PECOTA as a way to project the performances of pitchers and hitters, and it proved to be successful. Eventually, he moved into predicting the performance of minor league players, a much harder task because there was no base to build on. Between 2006 and 2011, PECOTA performed slightly worse than predictions made by traditional scouts, proving that scouts' judgment sometimes does have benefits that can't be derived simply from statistics.

The scouts had the benefit of mixing some stats with information that the system couldn't include. There are unquantifiable characteristics that also indicate whether a minor league player is ready to

## SUMMARY AND ANALYSIS

move up. The key to a good forecast is including as much information as possible, even qualitative information that must be translated to a data set.

Dustin Pedroia didn't seem like a natural choice for scouts because he didn't fit into a standard template for a successful baseball player, but PECOTA could see that some of what scouts saw as flaws—like his short stature—could actually be benefits for his position at second base. The Red Sox stuck with him during a rough first couple of seasons because the stats were showing that he had potential and that things could turn around. He also has an immense amount of confidence. He never doubted his own skills, and that focused perseverance was an incalculable benefit.

**Need to Know:** Nate Silver's baseball predicting system built on past models to become a successful forecast in Major League Baseball, but the qualitative information that scouts could incorporate made them more successful at predicting success out of the minor leagues.

## 4. For Years You've Been Telling Us That Rain Is Green

When Hurricane Katrina hit New Orleans in 2005, the National Hurricane Center made excellent predictions about the severity and timing of the storm.

Despite that, one-fifth of the city's population chose not to evacuate, mostly because they didn't think the storm would be as bad as it was. Sixteen hundred New Orleanians died.

Weather forecasting, broadly speaking, grew out of the Enlightenment belief that man could understand nature well enough to predict the future. Early twentieth-century weather predictions—both hand-calculated and those generated by early computers—were inaccurate, but weather forecasting is improving slowly, even as computing power grows at astronomical rates.

In 1972, Edward Lorenz coined the idea of the butterfly effect when trying to predict weather with an early computer. He and his team realized that even a tiny change, such as the number of decimal places used to represent a data point, could cause major differences in the forecasts. This led to his chaos theory, which occurs with nonlinear systems that are dynamic, meaning that the effect of one point affects its behavior at another point.

This is true of weather, making it highly vulnerable to inaccuracies. Inaccuracies can be caused by human error, such as our inability to record our surroundings with extreme precision.

As a result, weather forecasters run models with slightly tweaked information and evaluate all the different potential outcomes with these various data

## SUMMARY AND ANALYSIS

sets. For instance, if a weather forecaster says there's a 40% chance of rain, he may have run a variety of simulations and four out of ten times it rained.

At the National Weather Service, meteorologists improve on the statistical models with human intelligence, editing maps to include information that skilled forecasters can see but computers can't decipher. Predictions with human input have been consistently shown to be more accurate.

All the data produced by the National Weather Service is free to reuse. Private companies like the Weather Channel use this data to produce more consumer-friendly forecasts. For-profit forecasters have been known to fudge the data slightly for commercial reasons. They have what is called a wet bias, which means they overestimate rain: When they say there's a 20% chance, it only rains about 5% of the time. This is a deliberate choice, as people prefer to be surprised when it doesn't rain rather than surprised when it does.

Local news has even more of a wet bias, consistently overpredicting precipitation. This allows for more dramatic television, and news stations seem less interested in producing good forecasts, since people don't trust forecasts much to begin with. This circular logic is mostly harmless, but it can become crucial during emergencies like Hurricane Katrina, when viewers don't take the predictions of major weather seriously.

**Need to Know:** Weather forecasting, especially hurricane prediction, has become increasingly accurate in recent decades, but commercial incentives have meant that consumers don't always receive the most-accurate forecasts.

## 5. Desperately Seeking Signal

L'Aquila, Italy, had suffered several small earthquakes in the spring of 2009 before a 6.3 magnitude quake killed three hundred people and caused more than $16 billion in damages. Despite sitting very near a fault line, the city had grown complacent about the likelihood of serious quakes and was not sufficiently prepared.

In seismology, the distinction between a prediction and a forecast is important. A prediction would determine a specific time and place a quake will strike; according to the US Geological Survey, earthquakes are impossible to predict. A forecast, on the other hand, is a probabilistic statement about the likelihood of an earthquake in a certain region over a certain period of time. Earthquakes can be forecasted.

There are more than a million very small earthquakes every year, but very few large ones. When graphed logarithmically, earthquakes are strikingly constant—a magnitude 6.0 earthquake occurs ten times more frequently than magnitude 7.0 and one

## SUMMARY AND ANALYSIS

hundred times more frequently than a magnitude 8.0. So while it's possible to predict how often an area should get a catastrophic earthquake, there's no way to say *when* it might, which makes it difficult to prepare for.

What seismologists are struggling with is overfitting, the act of mistaking noise for a signal. This happens when a limited sample set produces incredibly specific conclusions that cannot be applied to more broad data sets. Overfitting is tempting because it makes a model look more accurate, when it in fact performs less accurately. This was the case in the models in Japan before the earthquakes in 2011—they had taken a slight over-occurrence of 7.5 magnitude quakes to indicate that this was as high as quakes were likely to get, and that a higher quake (like the 9.1 that they eventually got) was virtually impossible.

Given how rare major earthquakes are, it will be centuries before we can even conceive of using past events to predict future earthquakes. Some physicists believe earthquakes subscribe to the theory of complexity, which simply says that there are so many intertwined processes happening at once that we would never be able to predict them with any true certainty.

**Need to Know:** Earthquakes are possible to forecast probabilistically—how likely a certain type of earthquake is to hit an area in a span of time—but it is, at this point, impossible to predict the exact time or place with much accuracy.

## 6. How to Drown in Three Feet of Water

Political polls always include a reference to a margin of error, but economic forecasts rarely include such caveats. This gives the impression that they are extraordinarily accurate when they are not. When polled in November 2007, economists in the Survey of Professional Forecasters were confident that no recession was coming; they expected the economy to grow slightly in 2008.

In reality, GDP shrank by 3.3%, a likelihood that they said had a 1-in-500 chance of happening. On average, economists' GDP forecast have a margin of error of plus or minus 3.2%. But despite long-term inaccuracies, economists have not worked to improve their forecasts and reduce their bias towards overconfidence.

Part of the problem is that it's very difficult to determine cause and effect from economic data, which produces millions of data points. With so much data, it can be hard to understand the relationships between points. At the same time, the economy changes so quickly that indicators that are useful one year might be irrelevant only a few years later.

## SUMMARY AND ANALYSIS

It's also often impossible to distinguish correlation and causation, with factors like unemployment and consumer confidence sometimes leading as indicators and sometimes following. At the same time, economic policy put into place by governments affects the economy, meaning that past data about economic booms and busts must take into account the policies of the time.

One possible reason that the Fed's economic forecasts in 2007 did not predict a recession may be that they were focusing on data from 1986 to 2006, a period of very little volatility. They were ignoring data from eras with recessions and focusing on their success in forecasting during this calmer period.

Another factor is that economic data is subject to revision by the government—for example, in the last quarter of 2008, the government estimated the GDP was declining by 3.8% when it was actually declining by almost 9%. The data that economists are working with to make forecasts is often inaccurate.

In many ways, economic predictions and weather predictions suffer from some of the same problems. Both are dealing with dynamic systems, in which every small occurrence can affect every outcome, and both struggle to establish initial conditions on which to base forecasts. But weather predictions have been getting better because, fundamentally, weather obeys the same properties, while the economy is affected by human behavior, which is unpredictable and erratic. Human behavior should be considered when making forecasts, and a good forecast should be able to tell a story about the economy, rather than just following a bunch of data.

There's also a market for inaccurate forecasts, because partisan audiences want to hear a certain political forecast even if it's implausible. One solution might be to create a supply-side demand for accurate forecasts, such as a prediction market, where people make bets on the outcomes of particular economic or policy decisions, thereby incentivizing the investors to make or seek out accurate forecasts. A demand-side approach would require ignoring the more fantastical forecasters and focusing on the more responsible, perhaps including margins of error in GDP estimates.

**Need to Know:** Economic forecasting has a poor reputation for many reasons. Some failed predictions are due to personal motives—forecasters don't want to include margins of error in their predictions out of embarrassment—while others may be incentivized. Economic factors also change wildly over time, are hard to read accurately, and can cause drastic causation/correlation mistakes in forecasts.

**SUMMARY AND ANALYSIS**

## 7. Role Models

In 1976, soldiers at Fort Dix contracted the H1N1 flu virus, or swine flu, and one private ultimately died. This launched a massive panic: It was H1N1 that had caused fifty million people to die in the Spanish flu pandemic between 1918 and 1920. The US government spent $180 million producing a vaccine for the swine flu despite indications that a pandemic was unlikely. The vaccine they produced proved to be dangerous, causing a neurological condition known as Guillain–Barré syndrome. No cases of swine flu were reported the following year. In 2009, the virus returned. Again there were fears that a massive outbreak could lead to a global pandemic, and while the disease did spread around the world, deaths were rare.

Such missed forecasts are a result of infectious diseases being very hard to predict. When it comes to infectious diseases, early data is limited and unreliable, making it difficult to base predictions on it. And sometimes, predictions can be self-fulfilling: Media attention paid to a certain disease or disorder encourages people to identify symptoms of that disease more often and doctors to diagnose it more often. In cases like the flu, this can be good, because it encourages more people to go to the doctor and improves our ability to understand the scale of the disease, but it could also lead to people looking for symptoms of an illness they do not have.

There's also the phenomenon of the self-canceling prediction. This means that flu predictions could be inaccurate because the act of the prediction itself has led more people to make choices that prevent getting the flu.

The basic mathematical model for infectious disease is called the SIR model, which stands for susceptible, infectious, and recovered, the general path of a person's exposure to a disease such as the flu. A vaccine allows a person to skip the infectious stage, going straight from the susceptible to recovered.

The problem with this model is that it doesn't take into account that different members of the population are have varying levels of susceptibility, that some people are more likely to be vaccinated than others, and that it's rare for a person to intermingle equally with all people in their community.

One remedy for this flawed model would be the employment of an agent-based model, which would seek to simulate entire communities by accounting for, say, every single person in a city—where each person lives, where he works, whom he interacts with, and other behaviors that are consistent with his lifestyle.

These models can take into account human behavior to see how a disease ripples through a commu-

**SUMMARY AND ANALYSIS**

nity. The problem with these models is a lack of data; once a model is developed, its researchers must wait for future incidences of this occurrence to properly evaluate it.

Epidemiologists understand the limits of their data and their models and, therefore, are not overconfident in their forecasts. They see models as useful tools to test hypotheses and discover trends that could lead to useful action.

**Need to Know:** Forecasting infectious diseases is incredibly difficult and has not been very accurate. The potential of flu pandemics is often overblown by government and the media, while simplistic models don't account for the variety of human behaviors and the way that predictions can be either self-fulfilling or self-canceling. Ultimately, models built to measure the spread of infectious diseases are often best for identifying smaller, solvable issues within pandemics, rather than attempting to accurately forecast how a disease will spread.

## 8. Less and Less and Less Wrong

Haralabos "Bob" Voulgaris is a successful professional gambler in Los Angeles who spends eight months of the year watching dozens of NBA games every week. His success comes from analyzing information to discern meaningful relationships within data. His first bet was on the Los Angeles Lakers, when Las Vegas odds gave the Lakers a 13% chance of winning the championship, while Voulgaris thought it was closer to 25%. His $80,000 bet turned into a half million dollars.

Now that Voulgaris is very wealthy, he's constantly making bets on games. He focuses on small pieces of information that could have an effect—tweets from players, coded language from the coaches, press conferences, and his own observations from watching almost every NBA game—looking for meaningful patterns.

Voulgaris uses Bayesian reasoning, which argues that the more we learn about the universe, the closer we get to knowing the truth, even though we will never know for certain. Thus, we can begin to make probabilistic predictions based on our information that account for uncertainty but bring us closer to the truth.

Bayes' theorem assesses the probability of something happening by taking into account both the prior probability of its occurrence and the new event that has caused this assessment. Silver gives the example of a woman coming home to find another woman's underwear in her dresser and trying to determine whether her partner is cheating on her.

Using Bayes' theorem, he assesses the prior probability that the partner was cheating—in this case,

## SUMMARY AND ANALYSIS

using the statistic that 4% of married people cheat on their partners in any given year—while also assessing two factors related to the discovery of the underwear. One is the probability of the underwear being there if he *is* cheating, which Silver determines is approximately 50%. The other is the probability that the underwear would appear if he is *not* cheating, which Silver rates as 5%. Using Bayes' theorem and these numbers, Silver says the probability that the woman's partner is cheating on her is 29%.

Bayes' theorem encourages continuously updating probability estimates as new evidence presents itself. If another pair of underwear was found in the future, the woman would begin her prior probability of being cheated with 29%, thus upping the likelihood of the outcome of the new equation.

When Voulgaris is determining what to bet on, he's using the scientific method—observing a phenomenon, developing a hypothesis, formulating a prediction, and then testing that connection. For him, using the Bayesian idea of prior probability can affect how much he bets and how quickly he puts that into action. Silver believes that Bayesian prior probability is the key to moving closer to objective truth.

**Need to Know:** Silver introduces the idea of Bayesian thinking, which assesses the researcher's beliefs at the beginning of an experiment and constantly updates the experiment as new information becomes available. He uses the example of a successful basketball gambler who uses enormous amounts of data to look for meaningful relationships to beat the Vegas odds.

## 9. Rage Against the Machines

This chapter begins with a discussion of chess-playing computers, from the fraudulent Mechanical Turk of the eighteenth and nineteenth centuries to modern computer chess players, including IBM's Deep Blue, which beat Russian grandmaster Garry Kasparov in the 1990s. Chess poses an interesting computing matchup when it comes to technology: Computers have the benefit of making many calculations very quickly and remaining in perfect analysis for entire game, while humans have the ability to be flexible and learn.

Despite having all of the information about the rules and information of the game of chess, the human mind is unable to fully process all of the different possibilities of any game and therefore relies on simplified models, known as heuristics. Heuristics involve using rules of thumb to make a good decision when we cannot fully assess all the information to come to a perfect conclusion.

While the rules of chess are fairly simple and very well defined, there are an almost infinite number of

## SUMMARY AND ANALYSIS

possibilities in a chess game—in only three full turns, there are more than nine million possible sequences. The key for human and computer players is to break down the game into a series of smaller goals. For computers, the beginning is the hardest because it is a more abstract goal, while humans are better at understanding heuristics that can guide their decision-making. Midway through a chess game, there are billions of possibilities, which can be assessed fairly quickly by computer but not by a human, who must learn from past experience and balance the need for broad analysis with the need for deep investigation.

Deep Blue had been specifically designed to beat Kasparov by learning his traditions and by teaching the computer to play more like a human making tactical moves.

In the end, computers are good at things like weather forecasting and chess playing, but not at things like economic or earthquake forecasting, because computers can only work with what it is given. If computers have bad or incomplete information, they can't help. But if, as in many circumstances, you have some information but not a perfect understanding, the best thing to do with the computer is trial and error. This is what Google does, constantly testing everything from search result answers to the colors of advertisements to determine what is best for its consumers. It's a fundamentally Bayesian process, constantly refining algorithms, incorporating new data, and working toward some final product that may never actually be reached. Computers are only ever capable of processing information that we humans give them, and it is impossible to remove our biases from the models we create.

**Need to Know:** Though the IBM computer Deep Blue ultimately beat the grandmaster Garry Kasparov in a chess series, it had more to do with the information it was fed by its programmers to specifically beat Kasparov than it did with Kasparov's human desire to see patterns where none existed.

## 10. The Poker Bubble

The poker boom started in 2003, when an amateur named Chris Moneymaker won the World Series of Poker in Las Vegas and became an ESPN hero, convincing many people, including Silver, that winning big in poker was both easy and exciting. It was helped along by Internet poker, which was booming in the mid-2000s.

Silver began playing online, depositing $100 of his own money and quickly winning big. After six months, he had made $15,000 and quit his job. Though he didn't have a ton of experience, his statistical background gave him an advantage, because

## SUMMARY AND ANALYSIS

poker is mostly a mathematical game that depends on probabilistic judgments amidst uncertainty.

In Texas Hold'em, the most common variety of poker play today, there are 1,326 possible hand combinations dealt to each player, and a good player will make several hypotheses about what his opponent may have. Ultimately, this is an example of Bayesian reasoning, constantly updating the probabilities based on new information—in poker, that means things like checks and calls.

Because it is impossible to know what cards your opponent has been dealt, a poker player can only make decisions based on the information available, a factor often overlooked in television poker, where all of the players' hands are known and the focus is much more on the outcome. One strategy is to make your own play unpredictable, thus screwing up your opponents' probability calculations. This worked for Silver during his betting days, earning him about $400,000 from poker alone.

Part of his success came from the increase of bad players, known as fish, during the boom. Poker abides by the Pareto Principle of Prediction, also known as the 80-20 rule. In graph form, this shows a nonlinear curve where the first 20% of the effort results in 80% accuracy, but the following 80% effort only begins to approach 100% accuracy very slowly. In poker, this essentially means that learning the very basics will allow someone to make the same decisions as the best poker players 80% of the time, having only done 20% of the work.

But soon after hitting this point, there are diminishing returns, and you will suddenly face exceptionally skilled people who have put in an enormous amount of effort for a statistically small amount of improvement. This line is known as the water level, the threshold one must meet to even compete, and even tiny advantages can make all the difference. When the poker scene in the mid-2000s with was flooded with new players, it kept the water level low.

This all changed after 2006, when Congress passed a bill that made gambling online much more difficult for Americans. Professional players kept going, but many amateurs dropped out. Silver, who had been winning big, lost more than $130,000 in less than a year when forced to play without poor players to buoy him. The water level had risen.

**Need to Know:** The poker boom of the mid-2000s allowed people like Silver, who had a basic understanding of probabilities, to succeed because the market was now flooded with poor players. When online gambling became harder to win, the difficulty level rose, throwing out skilled amateurs like Silver.

**SUMMARY AND ANALYSIS**

## 11. If You Can't Beat 'Em . . .

Stock market trading has grown rapidly in the twenty-first century. Back in the 1950s, the average share of stock was held for about six years, but by the 2000s, it was held for only six months. In some cases, stocks are being bought and sold in a microsecond because, increasingly, traders believe that they can out-predict the conventional wisdom and beat the market.

In Bayesian theory, any two people whose predictions don't match can do one of two things: They can either discuss their forecasts and the reasoning behind them and come to a conclusion, both adopting one person's prediction, or meet somewhere in the middle. But if each is convinced he is right, they both must bet on their own predictions.

Bayesian theory has much in common with Adam Smith's invisible hand, and in theory, the stock market should be a perfect forecaster, constantly incorporating new data from various investors' beliefs in different companies. The efficient-market hypothesis even believes that it is impossible to out-predict markets, though recent bubbles and busts have more or less disproven this theory.

Another way to disprove an efficient-market hypothesis would be to prove that someone could consistently beat the stock market. This is difficult, as fees for trading can overtake gains, and because if there is a clear pattern, it is likely to be recognized by other investors, and therefore any individual benefit will be lost.

In the long term, price-to-earnings (P/E) ratios can help investors make reliable predictions, but this generally takes decades to see. Short-term predictions are often inaccurate because the act of buying a stock affects the price of that stock, multiplied over millions of traders around the world. Nowadays, most active traders are very short-term thinkers, so they are not looking to long-term projections. In every scenario, it is better for the trader to buy rather than sell, regardless of what happens to the market, which is why many brokerage firms only downgrade a stock when it becomes obvious they must.

One other thing that's changed in the stock market is the increase in institutional investors instead of individual investors. There are circumstances where it makes sense for traders to lose money for their investors if it makes them look more in line with their competitors and reduces their chance of getting fired. This herd mentality may be why bubbles form in the first place, as it is now in everyone's best interest to keep the markets going up.

In reality, bubbles may be easier to detect than to burst, and the Bayesian idea that you should always be willing to bet on your predictions doesn't always work in the real-world trading markets.

**SUMMARY AND ANALYSIS**

One perspective of the stock market is that there are essentially two concurrent tracks: one, a signal that is generally tied to business growth and other, the noise that involves feedback loops and herd behavior. This can make them seem predictable over the long term, but unpredictable on a day-to-day basis and means that chaotic nonlinear events like bubbles are built into the system. Silver argues that we can learn to detect and create softer landings when bubbles burst if we seek out from more information and recognize when they are happening, which requires understanding that markets are fallible.

**Need to Know:** Some believe that the stock market is too unpredictable to beat, but Silver contends that there is enough irrationality made by human errors such as herd mentality that we are able to recognize and prepared for any events, if not entirely prevent them.

## 12. A Climate of Healthy Skepticism

In 1988, a NASA climatologist testified before the Senate Energy Committee that the greenhouse effect was having a noticeable effect on day-to-day weather. Since then, the discussion about climate change has been fierce and political, but Silver argues that there might be a healthy skepticism about climate predictions that has nothing to do with climate-change denial.

The basis of climate change is the greenhouse effect, which is the process by which atmospheric gases absorb solar energy, which keeps the earth warmer than it otherwise would be. Human civilization can only live in a relatively narrow band of temperatures, and if more atmospheric gases like carbon dioxide, methane, and ozone are in the atmosphere, more heat will be trapped.

In 1990, the United Nations' International Panel on Climate Change (IPCC) broke down the existence of the greenhouse effect and the likelihood that as human activities increase greenhouse gases, the earth's surface will warm. As we've moved away from the term greenhouse effect to global warming and now climate change, there's been more and more misinformation that casts doubt on the existence of this phenomenon. It's important to be skeptical and apply Bayesian reasoning to every scientific theory, without cherry-picking information for partisan gains.

There are three sources of skepticism in the debate about climate change. One is money spent by the fossil fuel industry, which has an obvious financial incentive to maintain the status quo when it comes to energy. The second is pure contrarianism, where some people choose to place themselves outside of the mainstream, especially when there is any noise that can make their

## SUMMARY AND ANALYSIS

argument seem plausible. But the third, and most important, type of skepticism is scientific. This generally comes down to the skepticism about the computer models that forecast climate change. One skeptic argues that the consensus among forecasters is as much to do with partisan bias as scientific fact, that the global warming problem is too complex to forecast with any possibility of accuracy, and that the forecasters are overconfident and not accounting for the inherent uncertainty of global climate.

The term *consensus* is often used incorrectly when discussing climate change; it is not meant to imply that there is unanimous agreement, only that the majority of people, having tested hypotheses and analyzed findings, generally agree. The consensus used by the IPCC is potentially vulnerable to groupthink and herd mentality, and even those who agree with the consensus about climate change worry that the models being used by the IPCC lack necessary diversity.

Some of the skepticism about climate forecasting comes from people who are more used to forecasting things like economics, where there is a limited amount of data and the full truth can never be accurately assessed. On the other hand, the amount of uncertainty in climate forecasting is unique; unlike in political forecasting, where there are many historical examples to draw from, there's only one Earth with one climate, and predictions must make jumps decades into the future.

One problem that scientists feel about the climate change debate is that political partisans are discussing things that have been thought accepted by the scientific community, when there are genuine uncertainties that are not being discussed. This is politically dangerous, because people prefer overconfident forecasters. In reality, scientists have a good shot of making progress on climate change; politics will only make it more difficult.

**Need to Know:** There is scientific consensus that the greenhouse effect is real and worsened by man-made emissions, but there is genuine uncertainty about the models we use to predict climate change.

## 13. What You Don't Know Can Hurt You

In hindsight, there were indications that the Japanese were planning to attack Pearl Harbor in 1941. But this is only easy to see in retrospect—at the time, there was a lot of noise confusing the signal. Part of the problem is that an attack on US soil is incredibly rare, and therefore considered improbable and not taken seriously as a possibility.

It is similar to Donald Rumsfeld's famous line about "known unknowns" and "unknown unknowns"; there

## SUMMARY AND ANALYSIS

are not only questions we don't know the answers to, but also questions we haven't even thought to ask.

As with Pearl Harbor, there were signals that pointed toward the September 11 attacks. There were indications that Al Qaeda was targeting the United States (and specifically the World Trade Center), that they were using planes as weapons, and that they had the ability to pull off large-scale attacks. In the 9/11 Commission Report, the most serious type of systemic failure was the failure of imagination. We simply were not expecting terrorist to behave in the way that they did, and so we didn't properly register behavior that indicated that this was going to happen.

Predicting terrorist attacks is most akin to predicting earthquakes; there's a lot of signal, and smaller ones can indicate that a bigger one is coming in the future, but it is hard to sort through all of the incidents. Once you define terrorism and can agree to a count of terrorist incidents, you see that most terrorist attacks produce few fatalities and that the vast majority of fatalities from terrorist attacks will come from a few very large attacks.

As with earthquakes, it is possible to determine the statistical probability of a large incident without actually determining information about when and where this incident will take place. The difference is that, theoretically, terror attacks can be stopped. In this comparison, 9/11 is the equivalent of a magnitude 8.0

earthquake. A magnitude 9.0-equivalent attack could kill millions of people, as would a nuclear weapon.

Even though the likelihood of a very large terror attack is very small, just one could potentially kill more people than all terrorist attacks in the last forty years. Focusing on these larger attacks, as they have done in Israel, could potentially save a lot of lives. No more than two hundred people have been killed in once incident in Israel since records began in 1979, which is statistically improbable. But with national security, we are constantly led astray by our own biases and blind spots.

**Need to Know:** Terrorist attacks are hard to predict because there is a huge amount of uncertainty. They are like earthquakes—the damage by a few large ones is more significant than that from many small ones. But the biggest things holding back our terror attack forecasts are our own biases and our lack of imagination to consider the improbable.

## Conclusion

In time, as human knowledge continues to expand, we will begin to better understand nature's signals and predict events that we cannot now. But the expansion of big data may make it harder to predict human behavior, as the noise may increase out of ratio to the

**SUMMARY AND ANALYSIS**

signal. The key will be recognizing the gap between what we know and what we don't know.

One key will be to use Bayesian reasoning to detect and incorporate uncertainty and view the world probabilistically. This means acknowledging that we often view the world heuristically, with broad approximations, which are often oversimplified into inaccuracy. By slowly adding new information, we can get over closer to the truth.

It is also important to incorporate Bayesian prior belief, the estimate of how likely an event is before we begin to assess the evidence. This allows us place new information within a context and recognize our biases. The only way to improve our forecasts is to constantly be making them and learning from them. It will allow us to not place too much focus on one new, seemingly noteworthy piece of information, nor to ignore information that doesn't fit with our current forecasts.

The perception of our ability to predict is most influenced by our recent successes or failures with prediction. But the key to improving is learning from those mistakes but not letting them influence our current behavior.

**Need to Know:** To improve our forecasts in all areas, it is important to apply Bayesian theory on a consistent basis.

# Direct Quotes and Analysis

*"The instinctual shortcut that we take when we have 'too much information' is to engage with it selectively, picking out the parts we like and ignoring the remainder, making allies with those who have made the same choices and enemies of the rest."*

With the Internet, the average person is now bombarded with an unmanageable amount of information, and it is human nature to gravitate toward the data that fits with our existing worldview. But this only leads toward sectarianism. It is, of course, far wiser to consider more than one data point when forming an opinion and to listen to other points of view.

## SUMMARY AND ANALYSIS

*"The numbers have no way of speaking for themselves. We speak for them. We imbue them with meaning."*

By the act of using raw data, we are filtering them through our own biases and experiences. Numbers can be deliberately manipulated or cherry-picked to tell a certain story; more benignly, even an attempt to present opinionated data must be processed through our brains. So beware of making a decision and then finding the right data to back it up; look at all the relevant information that you can.

*"Some of you may be uncomfortable with a premise that I have been hinting at and will now state explicitly: we can never make perfectly objective predictions. They will always be tainted by our subjective point of view."*

This is one of Silver's main points. Acknowledging that we are working toward an objective truth, rather than already possessing it, encourages us to incorporate new data and be open to changing our perceptions of the truth, rather than seeking data that affirms our own beliefs.

*"Greed and fear are volatile quantities, however, and the balance can get out of whack. When there is an excess of greed in the system, there is a bubble. When there is an excess of fear, there is a panic."*

Many things cause economic instability, but a key factor is the reactions people have to economic conditions. When there is a bubble, it is better to sell and have some fear; when there is a depression, it is better to buy and have some hope.

*"When we are making predictions, we need a balance between curiosity and skepticism."*

This is the sum of Silver's thesis and focuses again on Bayesian theory and human bias. He argues that, when it comes to the business of making predictions, it is unlikely that any person or program will always be right. It's critical to expand your data sets as often as possible—and to continually question what you think you know. The more "skeptical" you are, the better a chance you have to uncover the truth.

# Trivia

**1.** Nate Silver's website, FiveThirtyEight, refers to the number of electors in the United States electoral college. This corresponds to the 435 Representatives in the US House, 100 Senators in the US Senate, and 3 electors from the otherwise unrepresented District of Columbia. FiveThirtyEight was founded in 2008 to cover the presidential election but has since expanded to sports, culture, and other areas.

**2.** Silver was named one of *Time*'s 100 Most Influential People in the world. In Silver's blurb, baseball statistician Bill James explains Silver's work and FiveThirtyEight as analysis of "data for the

## SUMMARY AND ANALYSIS

underlying order of the universe that is depicted by those facts and statistics."

**3.** In the 2012 presidential election, Silver accurately predicted all fifty states and Washington, DC. In the preface to the paperback edition of *The Signal and the Noise*, Silver says that he received a congratulatory call from the White House and that comedian Jon Stewart referred to him as "lord and god of the algorithm."

**4.** For the first eighteen months that he wrote about using data to analyze politics, Silver wrote under the pseudonym "Poblano," originally for the blog *Daily Kos*, and later for FiveThirtyEight.

**5.** The use of the terms *signal* and *noise* in analysis was made prominent Roberta Wohlstetter's 1962 book *Pearl Harbor: Warning and Decision*, in which Wohlstetter discussed how intelligence operatives failed to see the stream of clues that could have predicted the attack on Pearl Harbor because there was so much other data.

**6.** According to Merriam-Webster, the origin of the word *psephology*—the study of elections—comes from the Greek word for *pebble*, "because pebbles were used by the ancient Greeks in voting."

**7.** As weather prediction has improved, the National Hurricane Center has put considerable effort into presenting their forecasts in ways that are easy to understand and convey the seriousness of the storms. But it's up to local officials to declare things like evacuation orders, and in cases like New Orleans before Hurricane Katrina, Mayor Ray Nagin delayed issuing a mandatory evacuation order. This caused confusion and doubt and encouraged many people to stay in the city when they shouldn't have.

**8.** Earthquakes have historically been susceptible to some of the worst predictions. Famous examples include failed predictions in Lima, Peru, in 1981, when it was predicted that the country would experience the worst earthquake in history (it never came), and predictions that after 2004's 9.2 earthquake in Sumatra, Indonesia, another large quake was unlikely (an 8.5 hit in 2007).

**9.** Despite the political controversy surrounding climate change, the science of the greenhouse effect is simple, and as early as 1897, scientists predicted that it could affect the Earth's climate.

**10.** One complicating factor in the discussion about climate change is that temperature does not

## SUMMARY AND ANALYSIS

increase regularly and incrementally over time, but swings up and down, even though it is increasing in the long term. Such elements encourage probabilism—Silver puts the chance that there will be no net warming over the course of any given decade at about 15%.

# What's That Word?

**Bayes' theorem:** Named for statistician Thomas Bayes, a theory concerned with conditional probability, or the probability that a hypothesis is true if a particular event has happened. The posterior probability equation reads $(xy)/[xy + z(1-x)]$, where $x$ refers to the prior probability of an hypothesis being true, $y$ refers the probability of a hypothesis being true based on the new event, and $z$ refers to the probability of a hypothesis being untrue based on that same event.

**Bayesian reasoning:** A method of evaluating possible outcomes through probabilistic predictions that are constantly updated, also named after British statistician Thomas Bayes.

**SUMMARY AND ANALYSIS**

**Big data:** The vast amounts of information and data points now available through technology and the belief that this data is the key to progress.

**Calibration:** The accuracy or inaccuracy of a probabilistic forecast.

**Chaos theory:** The outcome of a system that is deeply connected and exponential rather than linear.

**Chartist:** A person who claims to be able to predict the direction of stock prices solely on the basis of past statistical patterns.

**Climatology:** The study of climate and weather patterns, including long-term historical averages of conditions in a particular area at a particular time.

**Collateralized debt obligation:** A complex type of financial product that includes collections of mortgage debt that are supposed to be structured to minimize risk.

**Efficient-market hypothesis:** The stock-market theory that it is impossible to outperform markets under certain conditions.

**Forecast (seismology):** A probabilistic statement about the likelihood of an earthquake in a general area.

**Foxes:** The political scientist Philip Tetlock's term for people who believe in many small ideas and are inclined toward nuanced predictions.

**Frequentism:** A theory of statistics that suggests that the only errors or uncertainty in a statistical problem are a result of a sampling error.

**Greenhouse effect:** The process by which some atmospheric gases, primarily water vapor, carbon dioxide, methane, and ozone, absorb solar energy reflected from the earth's surface, making the earth warmer than it otherwise would be.

**Hedgehogs:** The political scientist Philip Tetlock's term for people who believe in big, governing principles and are inclined toward significant, unambiguous predictions.

**Heuristics:** A problem-solving approach that uses "rules of thumb" when finding a definitive answer is impractical or impossible.

**Initial condition uncertainty:** The short-term factors that impact the way we experience the climate, like the weather, which can complicate the greenhouse signal.

## SUMMARY AND ANALYSIS

**Leverage (financial):** Using borrowed money to finance an investment or bet on an outcome.

**Long Boom:** The period between 1947 and 1999 when GDP (gross domestic product) and job growth were positively correlated.

**Negative feedback:** In economics, when two related factors are negatively correlated; when one goes up, the other goes down.

**Out of sample:** When the data set used to make a prediction fails to include relevant information for this particular forecast.

**Overfitting:** Mistaking noise for signal by trying to force information to fit a forecast.

**PECOTA (Player Empirical Comparison and Optimization Test Algorithm):** Nate Silver's baseball prediction system, which originally analyzed pitchers and hitters but eventually expanded into predicting the success of minor league players.

**Persistence:** The assumption that the weather will be the same tomorrow as it was today.

**Positive feedback:** In economics, when two related

factors are positively correlated; they both rise or fall together.

**Prediction (seismology):** A definitive and specific statement about when and where an earthquake will take place.

**Price-to-earnings ratio:** The comparison between what a company has earned and how their stock has been valued.

**Risk:** The odds of winning, which are known at the beginning of a bet.

**Sabermetrics:** The systematic study of baseball, especially with statistics.

**Similarity score:** An assessment of the statistical similarities between two baseball players

**SIR model:** An oversimplified model for treating infectious diseases; SIR stands for *susceptible, infectious, recovered*.

**Texas Hold'em:** A card game in which two cards are dealt to each player and five community cards to the board, and each player tries to make the best five-card hand out of those seven cards; a version of poker played online or in person.

## SUMMARY AND ANALYSIS

**Tranches:** The different pools (or groups) in a collateralized debt obligation that are meant to offset one another and minimize risk for investors.

**Uncertainty:** A hard, if not impossible, risk to measure.

**Wet bias:** A weather forecaster's tendency to overpredict the chance of rain.

# Critical Response

- An Amazon Best Nonfiction Book of 2012
- A *New York Times* bestseller
- A 2013 Phi Beta Kappa Award in Science winner

"What Silver is doing here is playing the role of public statistician—bringing simple but powerful empirical methods to bear on a controversial policy question, and making the results accessible to anyone with a high-school level of numeracy." —Noam Scheiber, *The New York Times*

"The first thing to note about *The Signal and the Noise* is that it is modest—not lacking in confidence or pointlessly self-effacing, but calm and honest about

## SUMMARY AND ANALYSIS

the limits to what the author or anyone else can know about what is going to happen next." —Ruth Scurr, *The Guardian*

"The strength of the book lies in the abundance of relevant detail Silver provides about each field and his analysis of why predictions are generally much better in some fields than others." —John Allen Paulos, *The Washington Post*

"[Silver] makes a convincing case as to . . . how things such as the financial crisis might have been alleviated or even avoided if enough people had considered their assumptions more carefully and searched for the ways in which they might be wrong." —Alex Koppelman, *Los Angeles Times*

"'The Signal and the Noise' is a book about prediction, not politics. In the spirit of Nassim Nicholas Taleb's widely read 'The Black Swan', Mr Silver asserts that humans are overconfident in their predictive abilities, that they struggle to think in probabilistic terms and build models that do not allow for uncertainty."
—*The Economist*

# About the Author

Nate Silver is a statistician and the founder and editor in chief of FiveThirtyEight, a data journalism site focused on politics, sports, and culture. In 2003, Silver created Baseball Prospectus's PECOTA, a projection system that uses sabermetrics to forecast the success of baseball players in the major and minor leagues. In 2007, he began writing about political statistics under the pseudonym "Poblano."

Called the "Kurt Cobain of statistics" by the *Boston Globe*, Silver launched FiveThirtyEight to cover the 2008 presidential primaries, drawing attention for accurately predicting forty-nine out of fifty states in the presidential election. From 2010 to 2013, the site was housed under the *New York Times* domain; in 2013, it was acquired by ESPN, and has since grown and expanded its coverage.

# For Your Information

**Online**

"The 12 Coolest Things We Learned From Nate Silver's Brand New Book." BusinessInsider.com

"The Best Nonfiction Books for Fiction Readers." EarlyBirdBooks.com

"Debates, Politics, and Predictions: Separate the Signal from the Noise." Wired.com

"Do We Want to Believe the Numbers? A Q&A With Nate Silver." TheAtlantic.com

"Nate Silver Extended Interview with Trevor Noah, November 14, 2016." ComedyCentral.com

"New School University Commencement Address, Nate Silver." C-SPAN.org

**SUMMARY AND ANALYSIS**

"Nate Silver on the Launch of ESPN's New FiveThirtyEight, Burritos, and Being a Fox." NYMag.com

"'Signal' and 'Noise': Prediction as Art and Science." NPR.org

"What Nate Silver Gets Wrong." NewYorker.com

## Books

*How Not to Be Wrong: The Power of Mathematical Thinking* by Jordan Ellenberg

*A Mind for Numbers: How to Excel at Math and Science (Even If You Flunked Algebra)* by Barbara Oakley

*Our Mathematical Universe: My Quest for the Ultimate Nature of Reality* by Max Tegmark

*Predicting Presidential Elections and Other Things* by Ray Fair

*A Student's Guide to Economics* by Paul Heyne

*SuperFreakonomics: Global Cooling, Patriotic Prostitutes, and Why Suicide Bombers Should Buy Life Insurance* by Steven D. Levitt and Stephen J. Dubner

*The Undoing Project: A Friendship That Changed Our Minds* by Michael Lewis

# Bibliography

James, Bill. "Scientists & Thinkers: Nate Silver," *Time*, April 30, 2009. http://content.time.com/time/specials/packages/article/0,28804,1894410_1893209_1893477,00.html.

Malkiel, Burton G. "Telling Lies From Statistics," *The Wall Street Journal*, September 24, 2012. http://www.wsj.com/articles/SB10000872396390444554704577644031670158646.

Paulos, John Allen. "The Signal and the Noise," *The Washington Post*, November 9, 2012. http://www.washingtonpost.com/opinions/the-signal-and-the-noise-why-so-many-predictions-fail--but-some-dont-by-nate-silver/2012/11/09/620bf2d0-0671-11e2-a10c-fa5a255a9258_story.html.

**SUMMARY AND ANALYSIS**

"Psephology." Merriam-Webster.com, accessed December 21, 2016. https://www.merriam-webster.com/dictionary/psephology.

Scheiber, Noam. "Known Unknowns," *The New York Times*, November 2, 2012. http://www.nytimes.com/2012/11/04/books/review/the-signal-and-the-noise-by-nate-silver.html?_r=0.

Scurr, Ruth. "The Signal and the Noise by Nate Silver—Review," *The Guardian*, November 9, 2012. https://www.theguardian.com/books/2012/nov/09/signal-and-noise-nate-silver-review.

Yglesias, Matthew. "The Silver Fox," *The Slate Book Review*, October 5, 2012. http://www.slate.com/articles/business/books/2012/10/nate_silver_s_book_the_signal_and_the_noise_reviewed_.html.

**WORTH BOOKS**
SMART SUMMARIES

## So much to read, so little time?

Explore summaries of bestselling fiction and essential nonfiction books on a variety of subjects, including business, history, science, lifestyle, and much more.

Visit the store at
www.ebookstore.worthbooks.com

# MORE SMART SUMMARIES
## FROM WORTH BOOKS

## BUSINESS

# MORE SMART SUMMARIES
## FROM WORTH BOOKS

## POPULAR SCIENCE

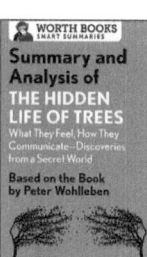

# MORE SMART SUMMARIES
## FROM WORTH BOOKS

## CURRENT AFFAIRS

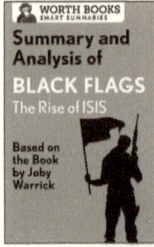
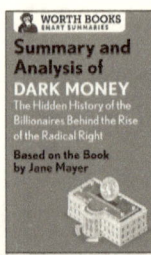

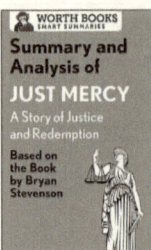
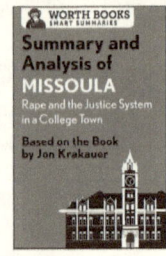
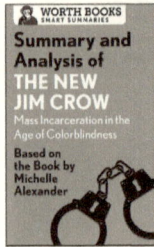
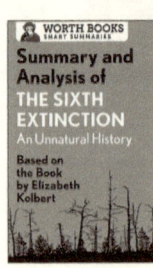

INTEGRATED MEDIA

Find a full list of our authors and
titles at www.openroadmedia.com

FOLLOW US
@OpenRoadMedia

www.ingramcontent.com/pod-product-compliance
Lightning Source LLC
Chambersburg PA
CBHW060342080526
44584CB00013B/879